BLOOM

MICHAEL LISTA

BLOOM

poems

ANANSI

Copyright © 2010 Michael Lista

All rights reserved. No part of this publication may be reproduced or transmitted in any form or by any means, electronic or mechanical, including photocopying, recording, or any information storage and retrieval system, without permission in writing from the publisher.

This edition published in 2010 by
House of Anansi Press Inc.
110 Spadina Avenue, Suite 801
Toronto, ON, M5V 2K4
Tel. 416-363-4343
Fax 416-363-1017
www.anansi.ca

Distributed in Canada by
HarperCollins Canada Ltd.
1995 Markham Road
Scarborough, ON, M1B 5M8
Toll free tel. 1-800-387-0117

Distributed in the United States by
Publishers Group West
1700 Fourth Street
Berkeley, CA 94710
Toll free tel. 1-800-788-3123

House of Anansi Press is committed to protecting our natural environment. As part of our efforts, this book is printed on paper that contains 100% post-consumer recycled fibres, is acid-free, and is processed chlorine-free.

14 13 12 11 10   1 2 3 4 5

LIBRARY AND ARCHIVES CANADA CATALOGUING IN PUBLICATION

Lista, Michael, 1983–
Bloom / Michael Lista.

Poems.
ISBN 978-0-88784-951-0

1. Slotin, Louis, 1910–1946—Poetry. 2. Manhattan Project (U.S.)—Poetry. I. Title.

PS8623.I85B46 2010   C811'.6   C2009-906399-9

Library of Congress Control Number: 2009939251

Cover design: Bill Douglas at The Bang
Text design and typesetting: Ingrid Paulson

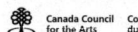 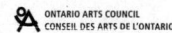

We acknowledge for their financial support of our publishing program the Canada Council for the Arts, the Ontario Arts Council, and the Government of Canada through the Canada Book Fund.

Printed and bound in Canada

for Elizabeth

CONTENTS

The Telemachiad
1

AM
3

PM
37

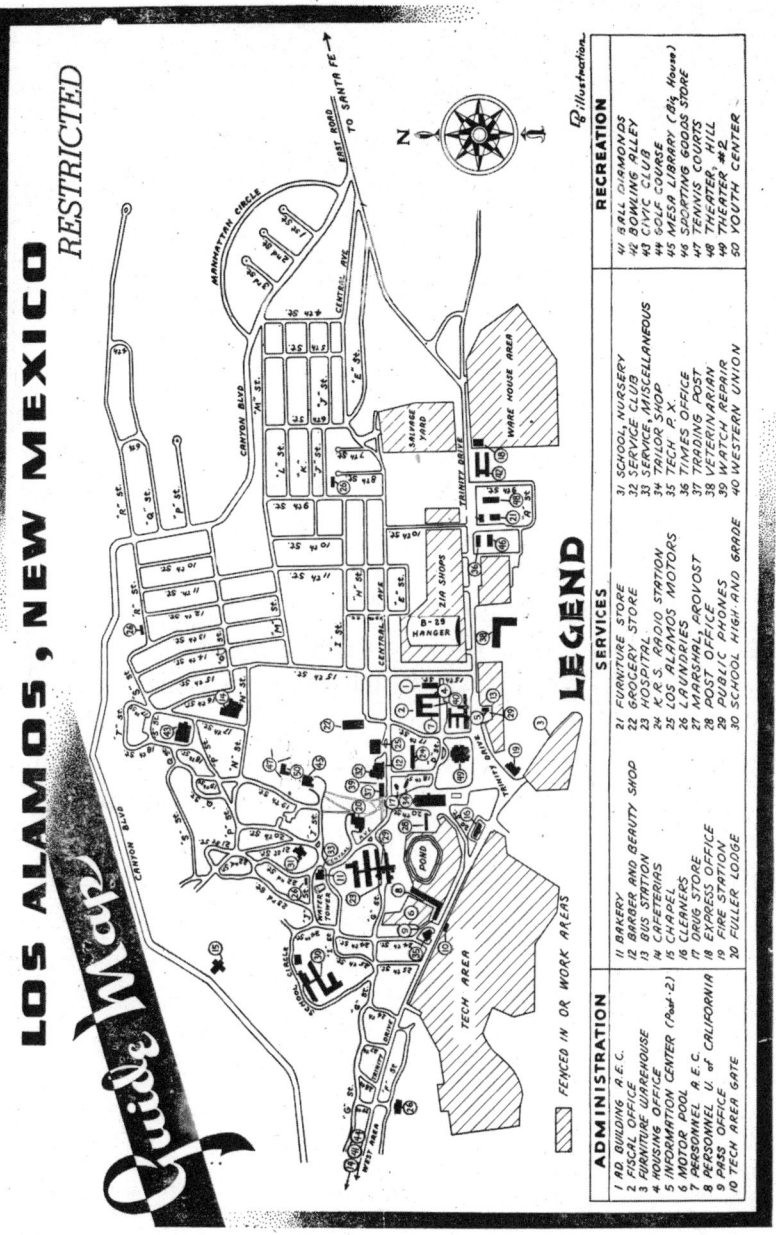

CREDIT: COURTESY LOS ALAMOS HISTORICAL MUSEUM PHOTO ARCHIVES.

And it was a sight worth seeing to behold the several souls choose their lives. And a piteous and a laughable and amazing sight it was also. The choice was mostly governed by what they had been accustomed to in their former life...

It so happened that the soul of Odysseus came forward to choose the very last of all. He remembered his former labours and had ceased from his ambition and so he spent a long time going round looking for the life of a private and obscure man. At last he found it lying about, ignored by everyone else; and when he saw it he took it gladly, and said that he would have made the same choice if the lot had fallen to him first.

Plato, The Republic

In the first, Harry K. Daghlian was working alone on the evening of 21 August 1945. Onto an almost complete assembly he inadvertently dropped a block of the tamper material used to reflect neutrons during an experiment scientists referred to as "tickling the dragon's tail." Criticality resulted at once. Daghlian died of acute radiation syndrome less than a month later. The Canadian Louis Slotin, the second victim, had won his doctorate in physics from the University of Chicago after serving with the Abraham Lincoln Brigade in the Spanish Civil War. At Los Alamos he became leader of the critical assemblies group. On 21 May 1946, working on the same "demon core" Daghlian had been nine months earlier, the screwdriver Slotin was using to prop one piece of reflective metal away from another slipped. A burst of radiation washed over all eight men in the room before Slotin was able to knock the piece away from the critical assembly. Slotin died nine days later, but his quick action saved the other seven men. Farther away and with the bloom partly eclipsed by Slotin's body, they were all over-exposed, but they all survived.

Barton C. Hacker, *The Dragon's Tail*

MOONRISE WITH POWER PLANT

An emerald halo glows beyond the night
Where the moon will rise hysterical and round;
It comes too slowly up without a sound
And swims above the city like a kite
That whitens through its realizing height—
Somehow insect men once walked around
On TV there, papery, profound,
Alive as moth wings at a summer light.

My soon-to-be-ex-neighbours start to fight
(He drinks too much, she isn't any fun).
It's not their fault. They do because they might;
What has the moon to do with being white?
I move tomorrow, following the sun.
It's hard to move in after anyone.

AM

RETURN OF BLOOM

LOUIS SLOTIN AND CALYPSO

At seven thirty-three my Louis is in the street and absolutely nothing of note is happening. Ok, ok, a car passes on his left and kicks up a little dust. He waves his hand in front of his face as if diffusing a fart. He shakes his head. It could be some of the dust kicked up by the car is in his face or it could be something else entirely. It's impossible for me to say for certain. Cicadas leap across the street, he crushes a couple. Where's his briefcase? Hold on—he's stopped, looking up at something in the sky above the base, an aeroplane maybe, or a bird, or some sort of vision of the Almighty. Sing to me muse. Nope, see, he's only sneezing, and there, after seven goes, he's on his way again. His shoulders rise a bit as he takes a happy breath. He could not eat for happiness. His black macintosh flaps in the wind. A girl rides towards him on her bicycle and she smiles as she passes, her thighs flashing. He tips his hat, the bastard. But it could be that the sun was in his face for a second, or he had an itch, or the wind picked up, who knows. Hell, he could have suddenly overheated. Just now he pauses at the foot of 3rd Street then turns the corner onto Canyon Boulevard.

A preternaturally large field mouse observes him, itches its ears, straightens on its haunches, as if in imitation, then scampers after him.

You should have seen his overcoat today. My favourite of his. It was long and thick, hand-tailored, either plaid or polka dot, or maybe, fuck it all, flannel.

after Daniil Kharms

METEMPSYCHOSIS

Beautifully Johanna Slotin slept
Till the sun laid its ladder on her head.
She woke hung over on her banded bed.
All this she could accept.

Graves was changing at the mirror
Into Louis's suit. "What time is Harry's thing?"
She asked, pulling and turning her ring.
He acted like he didn't hear her.

She went outside instead of trying.
Barefoot, in her nightgown, she walked her brillo lawn
—Yes, she'd get the paper—to where, below the dawn,
There laid a young coyote dying.

A transmogrifying mushroom
Sat giltrapped in a spunk of foam before his eyes
(Which rolled at her a dying thing's surprise),
A little bite taken from the bloom.

How quickly had the rot
Walked its mischief through the dog's dark state?
It stared at her. Now its hackles stood up straight
But the jackal did not.

A shine caught her eye—
A blue china doorknob lay unbroken by the curb.
She looked back to the dog (the morning now one perfect verb)—
It saw through her to the sky.

It had died.
She stood and breathed the rich air
Moving easily into her hair
And sighed.

after John Crowe Ransom

LOTUS EATERS

If we're forced to refer to one another, let's do it
the way heads of state or revered Rear Admirals reference
their rivals; as if we'd only known each other by reputation,
as a rumour in the other's classified reports,
or as a cardiac *blip* on a radar's clock-swept screen.
Let's do it the way the Enola Gay boys did—
with a wink, a wary tip of the hat, and a weary arc
across the night into the centre of the century.

And if you overhear people speak of our *split*,
listen for the schismatic overtones
of the word as applied to physics, specifically
fission: how when an atom's centre smashes and cracks,
new light explodes from matter's collapse,
a contagious backlash that shunts
neighbours from their paths,
exciting each moment, each moment exciting
further mutiny against its form
until finally, fully free, the gantry of our obsolescence
gives, and blooms its incandescence,
sucking breath from the chests of rubber-neckers.

after Troy Jollimore

LOUIS SLOTIN AS PIGEON FEEDER

Young man, you stick out like an ocean
around here. There hasn't been a cloud
in these here skies since the last Navajo rode free
through Los Alamos. The morning is as clear
as a sun-bleached bivalve. And there you stand
on the lab's sandy steps, in your black three-piece
and straw hat, drawing all the Brass's women over.
Is it safe to keep whistling them in
knowing their men have all the pistols
on the base? And why keep so obstinate a pace
after the general's slender daughter when she shies
from her suitors to the wind? If it wasn't for the bomb
you're building they'd call you a class-A moron
or talentless chauvinist, just another cowboy Ph.D.
with a pre-critical core at your disposal and a slotted flathead
in your pocket, predictably bragging to the spoken-fors
about tickling the dragon's tail alone in a lab room
of your own. But everyone saw you in the packed saloon
last night, introducing your bare-backed wife
to the sunburned, boozy lieutenants in their dress blues.
When your replacement, Alvin Graves, adhered himself to her
in the centre of the dance floor and blazed a finger
up the milky mile of her thigh you smiled, seemingly happy
to split your happy union like a helium nucleus, night

by night as the physicist in you sees fit. They thought it
was a stunt at first. Or a game that you and she play out in public
to achieve a private criticality of cells, to prime yourselves
for yourselves. When she and Graves paraded
through the charged miasma of the bar hand-in-hand
into the night, you didn't move, sat military still,
sipped your scotch and eyed back whatever women
watched you. Your mind lay open
like a drawer-full of forks. The staff and captains stayed late
to stoke what you had started and couldn't help but ask
Who are you putting on Louis
frisbeeing your wife out like a freebie?
What's in it for you when it's done?
How or who will end what you've begun?
This morning every eye attends you
like a brave new planet swum into their ken.
And now, fresh from their eclosion,
the base's women close in
on you, the thin stuff of their summer dresses
hurried into fast currents on their hips
by wind that stirs some birds into a mobile.
But be careful boy. Look around you.
The laws you're tempting to bend are firmer
than Fermi's. The rings fixed to their fingers will break
less easily than the valence of electrons, and will snap,
if at all, with much more force and violence. The science
that binds them will not abide your trying to annihilate it.

You know the clatter of their husbands' boots approaching
will scatter this flock, and two-by-two they'll tarry
to their quarters, leaving you with the sun to sink
between your knees as night encroaches, the desert
at your throat. There won't be time to pull your genius
over you like an overcoat as that night falls. Even now,
as the chapel bells cheerio, they break allegiance
with you. Like unstable isotopes they bloom, curio,
then fade into the everyday. Their pale bodies decompose
into the strobe of this New Mexican morning, as if by osmosis.
Still, the air flurries with the fragrance of their skin
as if wanting them was wanting's own neurosis.

after Peter Van Toorn

THE DANISH HISTORY OF SAXO GRAMMATICUS

The spared survive to misbehave,
The victors hoard water to spit:
Revenge is a dust they've
Each spotted in corners but permit

In the house of the soul to be
Stimulus for sneezes needed to annoy
Us. And on and on. It is we
Whom we ache to destroy

Us.

after Sean O'Brien

LOUIS SLOTIN OBSERVES
THE WIDOWED MINA DAGHLIAN

The balls on this chick!
After nine months tailing her
she shows up pregnant, sandals in hand.

Now I've got to nuke our city,
wander the smouldering caldera
and start my search for her again.

after Ladislav Skala

EN ROUTE TO DAGHLIAN'S MEMORIAL SERVICE, LOUIS SLOTIN READS *THE LOS ALAMOS TIMES*

May 21, 1946

State Police to "Get Tougher" on Roads.
Travel Policy to Change. Cancer
Drive Raises Hundreds, Spirits. Answer
Forthcoming re Windmill Building Codes.
Spike in Cases of Swollen Lymph Nodes.
Little Theatre's *Hamlet* Offers Laughs, Restraint.
2 Houses Get Test Coat of White Paint.
Despite Windy Spring, Banner Year for Toads.
Fire Care is Urged. USO Plans
More Plays Here. Lab Seeks "Trinitite."
Look Up: Complete Lunar Eclipse Tonight.
Scientists Pay Tribute to Brave Man's
Accident (continued on page three).
Poison Ivy Pretty; Leave It Be!

after A. M. Klein

AMLETH

Then his face went all red where it had been too white for a week and he said *I need to sit up a little help me up.* A fly lifted from his forehead as he spoke then it landed. The little worry in the wheel as they wheeled his body down the hall: and the screech on the tile as they skated the stern through the doorway. Harry was under the sheet and Johanna crying with her legs crossed and Oppenheimer helping the nurses get the gurney around the doorway. The O of his mouth under what white vowel. When would he smell? A month earlier he had strode into triage smoking, chuffed at the fuss, flirting with the nurses who buzzed around him gravely, buddying the doctors. And now he looked so small under the sheet, making his last noise down fluorescent halls. Never see him again. Death, I mean. Harry is dead. Harry Daghlian is nine months dead. He took my arm and needed not to be too tired to say this right and I blushed and his reddy yellow mellow smellow face needed to say this one thing right: *One life is all. One body. Do. But do.* His hand as soft as mist. Harry. He was Harry Krikor Daghlian. I hope he's somewhere he can piss on Nazi heads.

after James Joyce

LOUIS SLOTIN IN HADES

At chapel Harry is compared to Icarus,
Gluing government wings to too human
A back. The sea out ahead made of liquorice,
His foot on the ledge, a tug at the lumen
Of will and he's off. There below are the trees,
Standard as soldiers, beside him an ogle of birds.
Then he turns on his wing and revolves the world slowly
And he waves (his family wave back, tiny as words).
Go get 'em Harry! Over the roil and the blast
Of the ocean he goes, bulging like a sail.
I feel the heat at his face as he tunnels that vast
Eye — he points his finger — into what further vale,
Gone toward the light, buoyed on the air
By every evil thing to which I'm heir.

after Anne Sexton

A PINK WOOL KNITTED DRESS

In your thin linen summer dress
Before the earth had parched and cracked
You faced me at the altar. Bloomsday.

Sunshine—such that we squinted
Through that sequined Saturday
A decade in the making.
My suit—a pin-striped hand-me-down
From my father Israel's exhausted closet
He never knew I wore. But wore exclusively.
My shoes—sad, scuffed, crack-heeled
Penny loafers without a penny to them.

I was a wartime Jewish student from Winnipeg!
A dime to my name cost two to reclaim
So I borrowed. Not quite the frog-prince,
More tawny toad unchanged
Who bagged the princess anyway.

No ceremony could conspire
Against my inheritance, my whole wardrobe
Worn at once, my costume.
My wedding, like my work, insisted on secrecy.
But if we were to be married, you said,
It must be Queen of All Saints Basilica.
So under the marble dome we spoke

The words and were finally conjoined.
Your college roommate played bridesmaids
And witness and guests, even played groom's side,
Which would not abide a Catholic wedding
And so knew nothing of it. Slotins aside,
I hadn't even told a single friend
I'd stolen you. For best man we employed
An annoyed altar boy. His one condition:
He had to split before Eucharist
To make supper across Chicago for six.
But he stayed, and somewhere a table of Italians
Rose to a boil while we married.

                                  You were in bloom,
So tender and perfumed and thin,
A shuddering shoot of new tulip.
You shook, you clutched my sleeves and wept
With white joy, and I was the void
Over which the face of a god moved.
You said you felt the world unfurl inside you
As we stepped through the heavy doors
Into the west, sunshowers of fortune primed
To shine on us. And I'm on fire beside you,
Our hands empetalled, a bright new union
In the republic of light.

And every blossom trembled on its branch.
And every man and woman looked up weeping.
And every herd ran thunder through the earth.

                                          after Ted Hughes

LOUIS SLOTIN AND NAUSICAA

The room foreclosing, every eye on me,
I strut up to the sandwich table, poach
A tuna club, pour a draught of Beaujolais

And sidle up to three receptionists,
The voices as they're called, and me batting
My eyes to feel lovely. I feel lovely.

One shifts her weight onto her other leg
And stares across the Oriental air
Of Fuller Lodge, windows full of sky;

I think *you're pretty* but she doesn't blush.
The middle one—I dub thee Last Resort—
Holds out her hand. 'Ow you say *enchanté*.

I kiss her palm, I hover there too long,
My head absconds my shirt like a balloon,
My spine a pretty waving red valise

Synching off the asshole of my neck.
Mmmoppeh. When I right, I'm sweating like a jerk;
She pulls her hand away as from a stove

She knew was heating up but not how fast.
The last one knows me. Well. I lean
Into her ear and whisper "Baby eats

Spinach and prospers. Stroller on last legs."
Mrs. Graves giggles out a grunt
And says: "You have complete control of me

And you know it, you know it you bastard."
I chew some tuna, staring in her eye.

after Edna St. Vincent Millay

LOUIS SLOTIN AS THE WANDERER

Then comes evil over like a cloud.
Upon what mildhearted man it cast
In May, around him flowers dreaming of the day,
Overhead a desert bird widening its gyre,
That he should doff his bedthane's house,
Whose care and figure cannot hold him;
Lo, but ever goes that man
Through baffle gates, through cool unheated halls,
Full God drad that he the Wreaker
All mankind would fordo with fire,
Past ramparts of mesas, portcullis of sky,
Or lonely on fellows tway there chat
By trashcans, smoking by sedans,
He passes quietly, his mind unquiet.

There nods asleep, fatigued in the cloud-shade,
And dreams he of home, some man of Israel's folk
That wayfaring was stood by housedoor at night's oncoming,
His wife swiftseen at the window, waving him welcome;
Lightwise how her eyes kindled
Kissing her feet under a summer sheet.
And thitherto, in the dream from which he could not wake
Ongot he his reason:
His tribe was died, buried in lime,
And rightwise and willing was he
To outwander bellycrab on all who sorrow them.

Therefore you must not think to save him,
Though he will meet your eye, more familiar than ever;
You must not think on his house,
His waygone house where a woman waits,
Where the lightning blazes but is not loud;
Do not pray for his shelter; let him at last for to go as he came.
Do not come for him under the shadow of this cloud,
Its hour come round again.

after W. H. Auden

LOUIS SLOTIN'S SEX APPEAL

When our bespectacled physicist
bicycles the base, despondent past
the pandaemonic faces of his peers
    with their baby-eating smiles
and black fedoras shading
    their blackened brains
happily he thinks of his atomic bomb,
his happy Trojan horse,
his lips now breaking
    into the famous grin
that still dispatches women by the bed-load
into bedlam, into bed.

after Irving Layton

LOUIS SLOTIN ON HIROSHIMA

The moon was a slice
of sautéed onion the night
I decided I needed to die.

after Fitzpatrick Madrigali

SCYLLA & CHARYBDIS

It was a place to crash. Low on cash in June
Of Forty-Four and sick of spending Dad's, I was new
To Los Alamos and our barracks weren't built.
Our month-old marriage still a secret for our parents' sake,
The Project wouldn't pay for a place for us to stay,
Plus you had a couple courses left of your B.A.
Those were the days before the mesa boomed,
Blooming suburbs of blanched stucco
Where nothing else could ever grow.
Fresh off the bus from Santa Fe
And ready for a drink, I made my way
To Alexandra House. It had seen better days,
A sort of pre-V-Day VA for cheapskates,
A place to dash the night in peace and squalor.
The girls who helped run it were young war widows
Who lived above it for free.
I got a room on the top floor for cheap. It was tiny,
Had a window that overlooked I Street,
A bed that sagged in the centre, and a davenport
We'll call a carpenter's botched shot at cubism.
Mornings I'd report to coax the sun out of stone,
A miracle I was forbidden to disclose to you,
Laboured only at that and keeping close to you
Despite the redactors and distance.
Nights I'd squander all I couldn't care to save.
I wandered bravely into August

And started sharing my bed with a widow from Taos.
She was nothing like you, wore bolo ties,
Spoke easily of God and His language of surprise.
I think of it as a time I didn't own,
That I never had, so I can have it here on loan.
She and I slept naked each night as the moon
Ran its rote cycle around us into newness. We spooned
Though never once had sex.
The rub was keeping true to a covenant
We balanced between routine and accident.
She would walk her bent fingers over the dent
In my chest and though she'd linger there,
Her hand standing in a moonlit hollow of hair,
She never laid claim to it. She never dared
Distract my aim from you, so telescopic
That down the barrel of my days a myopic
Blur of you was all I saw. I was never tempted
To touch her except in brotherly comfort. Like siblings,
We never let our wonder wander from interest
Into incest. In September a new widow moved in with us.
A curt memo had informed her that her husband, Russ,
Had died in action against Japan. On the nightstand
She placed his picture next to yours. One hand
On the window sill, one hand holding a cigarette
Impossibly still, looking out, trim and pretty, in silhouette
Against the station lights. That's how I remember her.
Nights, she'd moan and beg me to enter her,
Weeping *please Louis*. I can't say with any certainty
How close my fidelity ever came to cruelty

And you will never know the war I waged
To ensure that we, the four of us, had a case for being saved.
I never caved. But with the world so hotly craved
By its own unmaking, I could see little good
In abstaining for the death that hovered like a hood
Above our heads. Lifting those naked, broken women
Through their early twenties into bed with me,
If they smiled, it took every suicidal atom in me
Not to see fidelity as resignation
To extinction.

after Ted Hughes

LAESTRYGONIANS

There are too many people on this planet
too hungry, locust-dumb, athwart each other, occluding the sun.
They leave behind a scorched field where they've loved.

after D. H. Lawrence

THE SPANISH TRAGEDY

Mina moaning down the hall
Through the Army hospital.
I hold her hand, encourage her—
She buckles as the cramps recur.
A doctor lifts her into bed,
Slowly lowers down her head,
Counts contractions by his watch,
Each tiny tock a tiny notch
Of knife—Agenbite of inwit.
Inwit's agenbite. It—
An hour in, in tears, in pain,
She picks a name: Louis Daghlian.
I turn to leave. She brays and moans.
My pestle name. Her mortar bones.

after Robert Frost

LOUIS SLOTIN TO HIS BABY BORN DEAD

Our stillborn daughter
Is about to stand up.
Is about to grow up.
Is about to lie down
And nap for an hour.
She lies around in the wreck
Of the room.
Breathe Molly breathe. What the heck?
She lies around with these flowers
In bloom.
How wild—
That unbreathing body was almost my child.
My daughter. Me in her eyes.
Let's go for a hike!
Her head on a pike
Has such red painted cheeks.
She is the sun ready to speak.
If you start to throw up,
If your heart starts to stop,
If your wife starts to mop,
If your voice starts to drop,
If your neck starts to flop,
Down in the ground within hours
Entombed.
Look at the world we've woken

Her into!
It's brand new!
She is about to leave the grocery store
With a bag as big as her body.
She's about to be naughty.
The rain is long-gone.
Tra-la.
Wake up!
Wake up, lazy bones!
Wake up!
Her voice is about to leave
Her clean sweet-smelling head.
Don't expect us to grieve
If you just lie there dead.

after Frederick Seidel

LOUIS SLOTIN AND THE WHITE LIE

I have never forgotten a day of my life,
belittled a loved one, made love to my wife
or sat in this chair in my office right here.
This isn't even my voice in your ear.

I couldn't put a face to my own name.
All the things assumed to be the same
hue they were yesterday change wavelength
as the sun ages, illuming us with less strength

than it did our fathers, our blood less red,
names of the living names of the dead.
When I stand between the light and the lightened
I draw a human black over the brightened.

The night the Little Boy fell on Japan, a
brighter me looked down to see Johanna
beaming up with such transparent wonder
she saw through me to worlds I'd set asunder,

all I had broken, a human city
named into amnesia. The opacity
we wore as marriage masks cast off,
the light between us dropped into a trough

as she turned away and there it stayed.
But consider this: things we've betrayed,
forgone, or loved too deeply to desire
are reenlightened in divorce's fire

just as if I parroted *bird, bird, bird*
at that starling in the tree, the word
would slough its subject like its subject's doom
and fall from meaning into meaning's tomb;

then, subjects ourselves, it would illustrate
we are the source, not what the source illuminates.
See your shadow flee a garden patch
as your namesake and your name detach

and you turn to glass. In that transparency
the Love That Lights would show no clemency
to such a solipsistic soul so isolated
and would explode in it a flash ignited

like detonated dawn which humbles lamps
still staving darkness off their human camps.
(And how unlikely and how right it is that night
must both dissolve and rhyme with light.)

But like a peerless, stationary God
without a face, adorning each façade,
no one at one with all the living realm
can love one thing whose death will underwhelm.

Only by this (goodbye sweetheart, goodbye)—
the opaque, plastic facts and the white lie—
do we keep the world, as we might beach
that interrupting dawn to keep its bleach

off all the stars there hanging on their hooks,
that everything could keep the tinted look
of life, our gods high up enough above
and our beloved far away enough to love.

after Don Paterson

SIRENS

First light tomorrow, before I leave
We'll meet. I promise. I'm coming
Over the mountain at dawn. Believe
You me. I'll crest the summit running.

Undistracted and silent, I'll come the way
A survivor would hurrying home
Uninterested in what to see or to say
Being the news of my being alone.

Empty car after empty car, my brain
Is a train to tomorrow, to the tomb
Marked *Harry Krikor Daghlian*
Where I'll lay holly and heather in bloom.

after Victor Hugo

PM

EXIT BLOOM

LOUIS SLOTIN NODS OFF

When a man is succumbing to sleep in a chair, his arms folded across himself, his chin to his chest, ankle over ankle, and beyond the porch a diesel engine keeps whatever time passes, then it seems to the man as if an immense metal dome is closing down around him, and that the thin, happy static of his thoughts is peeking out unworriedly as the jaws chomp down around his head, and he half expects, as his back relaxes like a fiddlehead fern, that his mind might fall agreeably to the deck with a bang, and roll off the lip into ruff, never to be found again.

It just so happens that a man by the name of Louis Slotin is at this very moment looking perfectly relaxed in just such a chair, on just such a porch, and he is nodding off. But as the dome lowers its dark around him, increasingly panicked thoughts, in the shape of a cornered coyote, peer out of his increasingly metallic head.

after Daniil Kharms

LOUIS SLOTIN'S FLAW

I wake to Graves cavorting with my wife.
They're at the outer gate. She looks upset,
she's pointing over here. A wind blows flat
and hard across them both. She grabs her hat
just as it lifts away, a gesture set
in all its causes since the knot of life

that ravels wholly out as history.
If there's free will, it's something like a hat,
set on my head, made for my head, though free
to blow clean off but then be caught by me.
One world, one motion, marshalled in combat,
its reflex rhymed against its destiny.

My wife and my replacement! Look, they fret
a pause then burst into a forced couplet
then quiet, strangers out of sync. My God
she's gorgeous. Her nose looks like a pig's!
She toes her Molly Janes at dead dry twigs,
draws in dust a line she dares the flawed

to cross. High noon. Even their shadows hide.
Last week as man and wife, a palimpsest,
we met there too, walked home, and went to bed.
She took her earrings off and then undressed,
peeled back the sheets, slid in, and kissed my head,
our hats hung on the coat rack side-by-side.

And now the wind picks up, his hat lifts
up and off, and he chases it
as the breeze uncoils, succors and shifts
like a genius toying with a lesser wit.
This little life we launch in winds askance
and skipper blindly to its last flawed chance.

The bastard knows I see him there.
He recovers, waves and will not scare.

after Robert Lowell

OXEN OF THE SUN

From Alpha Site to Omega Site,
It's a ten minute drive up the mountain,
Under canopy, candelabras of sunlight
Chiming the cool of a fountain.

Bandalier Monument. A thousand years
Since the Pueblo people called this home.
One day they left. An eye through its tears
Looks at last on a kitchen dome.

*The stone age to the nuclear age
In fifteen steps.* The lab's right there!
Same path, same feet, same rage
To stoke and quench, to dream, to dare—

I stumble in their barefoot wake.
The last must finish what the first begins;
To crush, to break each neighbour, take
Each life off like a coat, and hang their skins.

As a boy I wanted to grow up and hunt.
Got myself a meaner bow is all;
What is this dark engine in us that shunts
From age to age, as from rail to rail,

Picking up steam as it barrels on?
I've always followed it. Only today
The long black figure on the lawn,
Stubborn as a shadow, is yesterday.

after Seamus Heaney

WANDERING ROCKS

After a half-decade dedicated to halving lives—
each only ever half-there, and love a half-life
to be absent through and waited safely out
lest by it we be fatally inured, but never injured—
and after the Lancasters quiet, and we've scraped

the scrap evidence of supper from our plates, having
escaped the booby traps of dinner chat unscathed,
and instead simply complained about the heat
and lack of rain, again, agreeing, and after having
graduated to brown booze and reading glasses,
to cull the news of all the pain that we can claim,

do you think that we can ever dial back our names
to their static whole note, that cracked octave
all chord and no progression?

after Nick Laird

THE RETURN OF ODYSSEUS

The lab door swings. Observers saunter in.
The latches moan for every bedswerver,
Let cuckold, jerkoff, ass, and fuck-o through.
The laboratory and the halls resound
The stench and frenzy of a boxing match,
Men in black lean up against the walls,
Snuff their smokes to little periods—
Broken smoking bones litter the floor.
Some even spit, some eye Louis and smirk
And gossip as he lays out instruments.
Johnson farts and everybody laughs,
Giants, Yanks, without a law to bless them,
Everyone as easy and relaxed
As pledges in a swank fraternity.

Quiet in the centre of the room,
Slotin starts his solitary loom
On which the cross stitch of the quilted world
Was spared unravel by unravelling
To be unravelled at some later date
Or spared. Slotin thinks: "Here I put nothing.
Less than nothing. I stitch emptiness
Into the brutal knot of history.
Harry. Harry, this is what we've done,
Our gift, our grief: to keep a vacant space

Where some redeeming novelty may bloom.
We are the centre never meant to hold."

Where none should go, where all have been, he waits.
But even now, both Daghlian and Graves
Are on the long circular way to him.

after Edwin Muir

DO. BUT DO.

What is this weird sphere
I'm assembling?
It's a question better left unasked.

Say I'm making it
for making's sake, as humans must
when put before an erector set
whose pieces spell out
Please for the love of Jesus
do not dare assemble us
its finished frame a ciphered skeletal
you shouldn't have.

But it's a question better left unasked,
running one's fingers
over a cool hemisphere of plutonium.

Say I'm making it
for unmaking's sake, as humans must
when threatened with unmaking—
we must race ourselves to our end
or be beaten to it.
Even our unmaking is unmade
when the cut cloth of our death in the dirt
vanishes as the sun slides down beneath our knees.

Either way I assemble this weird sphere
because being human, able, and put before it here,
I must—just as my hands (come fresh
from her hips in the dazzle
of dusk descending on the dusty mesa) must
feel fingers-first down the darkened hallway
for the table upon which is a vase
of whatever's in bloom.

after Robyn Sarah

CYCLOPS

Alvin Graves closes the door and approaches Slotin.
    —Afternoon, gentlemen. I hear you have a torch in need of passing.
    Slotin waves him over:
—One last burn boyo and then I here abjure my magic. There's an art to starting a fire. There are creative arts and there are recreative arts. We'll call this one fun from scratch.
—Let's see if I can't pick it up.
—Not too much wood, Slotin says, forcing his thumb into the tamper's empty eye, that's the rub.
    Slotin levels the hemisphere to the bevel of the screwdriver and listens to the core crackle as it catches. Graves dissents tacitly:
    —Don't think you're supposed to string the bow like that.
    —I'm almosting it, says Slotin, man of many ways, winking a made eye. Nice suit, he says. Stand closer.

after James Joyce

LOUIS SLOTIN IMPROVISES

I'm beside Slotin
Who has the sun in one hand, a feather in the other—
The flash he tickles out giggles his name.
So he takes his zipper in one hand,
The Milky Way in the other—
The flash that fish-thumps out pronounces the alphabet,
In which every other letter is "I."
He takes Auschwitz in one hand
And Japan in the other—
The flash that gnarls out proposes to marry him.
So he takes his stillborn daughter in one hand
And his sense of humour in the other—
The flash that claps out detaches him from history.
So he takes his grandfather's skinned pelt in one hand,
And a smelted pickerel in the other—
The flash that cooks out spoils his ballot.
He grabs his wife's lover with one hand
And his passed kidney stone in the other—
The flash that flirts out confuses the tenses.
And so he takes his sister's fondness for hyperbole in one hand,
His sotto-voce in the other,
And the flash that booms out falls in love with him.
He holds his heart, pounding like an ache in one hand,
And with the other feels for a camera—
The flash that blooms out sings him an opera.

So he takes his birth-scream in one hand
And his death-please in the other
And lets the flash scorch him to dust.

And with a gesture any waiter would envy,
He pops the top on his creation,
Stands back, panting,
As if to say
Voila!

after Ted Hughes

THE ECLIPSE

It is here during the London premiere
Of *Fat Man and Little Boy*, 1989

(*Shadow Makers* in the UK) that John Cusack—
Who plays you Louis—rises in the back

Row and crosses god-sized across the projector,
Obliterating hugely what had promised to be

His name-making scene.
Every patron in every seat

Turns to the backlit star
And misses you in your name-making scene.

The soundtrack holds its breath.
The fountain sodas hold their breath

Then gasp bubbles as he swoops
The scene into shadow. Onscreen

(Borne on Cusack's back), Cusack screams
"Everybody back" then the screwdriver slips.

Zeus allows a loud lone crack. Someone shouts
"Critical." Someone shouts "Move."

Dustmotes orbit like a sky reordered—
He is Lazarus with a blue light bearing him on.

The panic passes as he trips accidentally past
The key grip, gripping a seatback

And the scene crests the east
Of his shoulder heroically.

He hears himself say "mark the positions
You're standing in then get out"

As he crosses the lobby into the street.
His cab noses home to the hotel.

In the backseat, he turns to the window
Where the moon is wearing its own face as a mask.

after Nobody

*B*ut *Noone* is there.
    Screams: Louis!
    Mortar blasts, warplanes, flares.
    Begs: Louis!
    Gunclatter, radiochatter, blueflash;
    Whispers: Louis!
    And again: Louis!
    Then comes the light to run the word away.
          * * *
But Noone is there.

after Rilke

LOUIS SLOTIN AND THE GREEN KNIGHT

The one-eyed knight kneels, hanging his heavy head
ahead of him, showing a shock of green skin
as he collects clumps of hair up his head
off his neck, now just a naked nape there,
a slice of salami sandwiched by steel.
Dr. Slotin grips his silver sword then hangs it high—
digs his left loafer into the floor,
then sends the scimitar slowly into the soft skin.
The strength of the swing splits the spine
like a line, flying through the fat so effortlessly
it nutellas through the neck and flobbles in the floorboards.
The huge, helmeted head eyes round the room
which dims while the shucked skull
glows green, and the physicists foot it.
Blood gushes like beaujolais from his jugular,
blooming Vesuvian at the base of his body armour,
but the knight neither faints nor falters nor flinches
rather slowly arises, standing on legs like sequoias,
then gropes his gloves around the ground,
feeling for facemask, which he finds at Graves' feet,
then holds it high, hurrying to his horse
tethered to the test table, mounting its back in a bound,
straight into the saddle, holding by the hair

his own decapitated—but quite undead—head.
Then he settles in his stirrups like a tourist in Taos
going galloping, not a man recently estranged
              from his brain.
    And when he rears at us
    the last blood left to drain
    pearls out thick and viscous,
    and Slotin has become Daghlian.

after the Pearl Poet

HAMLET

The air in the room was beginning to cool and goose nipples pimpled his arms. One by one they had left the room after Slotin, but Graves had stayed. He stepped cautiously toward the table upon which the screwdriver laid. But hell if he'd reach for it. Whatever other world lay etherized beyond them be damned; he would no longer speed boldly to it with the rest. He'd rather bear the ills he has than fly to inwit's agenbite. Agenbite of inwit. Misery. No—he thought then of Johanna whose eyes last night for all their excess brimmed nothing of him. Misery.

Graves began to cry. He had never felt this for a woman but he knew he was in love. The tears came in chokes and coughs and blotted out his eyes and the room dimmed and in the gathering darkness he saw Slotin's silhouette in relief against a bolt of blue where the bloom had been. Time is out of joint. The grass is singing, only the wind's gone home. His soul had intruded onto that vast updraft where rise the dark vessels of the dead. He was aware of, but could not fully understand, their flotsam, cycling drift. And then his sense of self dissolved into a black—he closed his eyes—anonymous leaf clinging to the black bough of the world, where once the dead had blossomed, waved, and reddened.

Red in the window made him turn. The moon's eclipse was almost complete. Sleepily he watched its face surprise into black. The time had come for him to leave this place. Yes, the forecasts had been correct: a complete eclipse was general over the West. Sleepily it darkened every part of the lush caldera where once the earth had seized and blown and reddened out the fingerlings of mesas, sleepily it darkened the way of highland elk, sleepily it darkened the deep wound of the Colorado, and westward, it darkened in the mutinous Pacific waves. It darkened, too, every part of the churchyard where Harry Daghlian lay buried. The light lifted from the plain headstones, rendering the names anonymous and all their dates, it lifted from the lacquer of the little gate, it lifted every shadow up, night a deeper night, darkening the darkened and the dead.

after James Joyce

LOUIS SLOTIN EXITS THE OFFICE

A sickle of pale-faced physicists
surrounds me as I vomit on the sand—
the open maw of the desert
yawns,
sick & tired of pronouncing me,
its last infected tooth loose & cavitied.
Across the Trinity Caldera, a steam train
to Taos sounds its sonar at nowhere. Ducked,
akimbo, palms on thighs, I say
You fine fellows should fare ok
but I think I'm fucked.

after Fitzpatrick Madrigali

ITHACA

In the bathroom, a mirror and a razor
crossed over a day-old bowl of lather.

A bed, unmade, a patio
in progress, KRS radio

reporting the eclipse, the butter-side
down of all Johanna's journal has to hide

like landed toast upon the table. A metal
knuckle of keys shucksing the kettle.

By means of apology, a parent-
to-be to parent-to-be statement

of fact on the counter: *need milk
Louis.* Her white silk

scarf—her *white elephant*—artfully coiled
into a question mark, indicting their spoiled

experiment from the kitchen floor.
An unremarkable hallway, a door

opened to their lawn, and her there, neither old
nor any longer young, in neither silver nor gold

but burnt into a single finger a ring
of unadulterated white, which means everything.

after Simon Armitage

HEAD OF A DANDELION

This is the dandelion with its head of identical thoughts.

An elderly gentleman grabbed by his shoulders,
Threshed bald and thrashed older,

This is the dusk-bloom shaken to husk.

This is amnesia blossomed as cotton,
Its life a blonde head its blonde head has forgotten,
Diseased downy and threadbare

Like a strawman had strolled a Niagara of flames…

Forever forever the same:
One thing coils hugely,

The other lowers its head, repeating its name.

One thing unweathers its effortless whelm,
Blows hairdos of trees, combs their weaknesses free,
It flanks shadow-mouthed, flashes oceans to tin;

The other thing gives, surrenders its skin,
Stops sweating, relaxes, goes soft as it's bitten…

And this is his flower.

His head is a mess of a meteor shower,
Each recalcitrant petal a wish past a window

Until a last eyelash winks free:
I'm not going anywhere——
      I'm going everywhere——

after Alice Oswald

LOUIS SLOTIN AND EUMAEUS

He was thirty-six years old when he arrived inside the hourglass.
We started to sink through a hole in the desert which hissed us in

Beyond its event horizon to an identical Los Alamos, upon which we landed
Unhurt but instantly amnesiac. As the sucking restarted we recalled

How many woozy eons we'd been sifting from instant to pseudangelian
Instant, siphoned past deck chairs and swineherds and Mameluke swords.

Slotin would be handling some black ball
When I'd smile and the suction would start.

When I realize how many times that has been said I said,
And that you have to grin through its retelling I want to disappear.

And then that everafter of inertia, as if reeling from an awkward anecdote,
And that whisper at our feet, like an orchestra of crickets snickering

Before we began to sink through another second hand-in-hand
Down that glass canal, as if being reincarnated as ourselves.

after Luke Kennard

CIRCE

Everybody in the town turns
As the aeroplane approaches.
Louis's in his baby's bed in the woods.
Coyotes cuddle up, tubes

In their wet, black noses.
He is awake. Hermes,
A big boy, is about to climb
Into an itsy bitsy bed.

Suddenly tree-shadows dial
As Louis turns on the light.
Everybody in the town stops.
The world isn't only what it is. They cower

Then puff. Sweet
For his sweetie, Louis puts
Moly flowers on a cool rock
For his princess, tubes

In her nose. Louis approaches a tree
And its leaves colour
And fall. As he passes
It blooms. Everybody in the town

Is a bad little piggy and he loves them.

after Frederick Seidel

LOUIS SLOTIN'S GOT THE MAIN GATE BLUES

Someone has given Louis the spare key to Paradise.
Here, he said, take it; it won't be offered twice.
High-five the dragon, lay beneath the tree and grieve,
Hell, chloroform Adam and make a pass at Eve.

Slotin walks up to the fence. It reads "Electrified"
And "Guard dog on duty." Below, an undignified
Bulldog with wings sleeps. Slotin has to sit.
He's holding the holy key, and he's afraid of it.

after George Johnston

METEMPSYCHOSIS

He is walking down your street towards you.
His heels clack uniquely down the night street. Your lights
are on. He's been watching since you slinked
before the sink, ran the faucet, started
washing dishes, how your shoulders shake
like something itches. You keep your back turned.
How you know he likes it when you act
like you don't know he's looking. The door's
unlocked. You've put on something nice.
And the breeze he's combing through.
And the breeze—but now you smell it isn't he
who's being blown to you. Amiss comes dressed
as what you hadn't heart to fear, and now
disaster is an accident arrived,
prescribed, a doorbell rung just as the record's needle
lifts, and the crackle fades in key. As if your gifts
are what (you realize)
you'd wrapped. Your husband is a bit of news
your lover will deliver. Only facts remain,
a laundry of gestures he will hang with care
that you'll be left to fold. He shuts the door, removes his hat,
and steps to you. You keep your back turned
how you know he likes it. Because in turning,
this sink, however he looks,

the old star-eaten blanket of the night
will hook its hinge, as the freight of what must be
assumes its weight. To see Graves speak,
the latch of his tongue taking its lock. To turn
and face your fate, a name in its mouth.

after Karen Solie

JOHANNA FINDS A REASON

Yes baby I believe
In everything you say;
Tomorrow will relieve
Us of what came today.

after The Velvet Underground

THE COMING OF WISDOM WITH TIME

Though years are many, the day is one;
And all the novel shades of history
Are gnomons dialled by a daily sun.
Now let me set into the mystery.

after W. B. Yeats

LOUIS SLOTIN YOU WILL NOT TURN FORTY

At your fortieth birthday, on a moonlit beach,
One of your guests is late.
You save a plate. A place is clean and set
Amid the after-dinner mess.
Why are you upset? Everyone is here

And laughing wildly, the whole extended gang—
Old friends and family, your little niece,
Your coterie of Nobel laureates, some heads of state,
And Harry's here, and Alvin Graves, their wives,
Their eyes run over with ecstatic tears—even

The palm trees rock with laughter, tossing their hair.
Your forty candles come tickled
To the table, trembling with joy. And your daughter
Laughs from your lap. Japan laughs deep through its ash.
Johanna puts her arms around you and the sky

Turns like a Ferris wheel, the stars
Smiling from their spokes.
But your bomb—
What about your bomb?
It's the only one missing!

We wait all through the night for it to come.

after Ted Hughes

LOUIS SLOTIN CIRCLES KILIMANJARO

They are lowering him onto the backbench of an idling car when the jackal starts in, amused, almost human from the black beyond the headlamps. Two sifting clouds of insects blizzard in the light, twin portholes to the bottom of the sea. The men hear it and turn as it sickens into pitch and out again. Slotin is asleep and doesn't stir. In his dream he is at the house on Scotia Street in Winnipeg the night before he is to leave. His mother gives him a sword, his father tells jokes he doesn't understand. Somehow Johanna is there they have been quarrelling and in the kitchen he says Mum and Dad this is her but her back is turned. She says we only fight when we feel our finest why is that and how cruel he'd been and how hungry and how *odd is he* to be anything at all. When is it? May the twenty-second? Sure, the blooming things is all over. Then the jackal screams so loud it shakes the windows and his hand is the shape of a cantaloupe and Johanna is calling.

"Louis," she calls, "Louis! Louis!" But he is asleep he is weary he has travelled and the rear red lights dim and the car lurches forward.

Then she screams, "Oppenheimer stop!" Then her voice brittles to its cracking point "Goddamn it Oppenheimer stop!"

But the car does not stop and she can see her husband's head slump and bob and see his hair ruffle in the wind and she opens into flat sprint the taste of pennies in her mouth and the jackal unwinds its siren but she does not hear it for the blaring of her heart.

after Ernest Hemingway

PENELOPE

solemn *declaration*	Lift him. With a voice of waves. Capitals report flocks of birds low between towers. In sensitive individuals, strange cough, homesickness, déjà vu.
nuclear war *is unthinkable* *threshold*	Detected by stairway to the stars. Bananas peel. Grapes deflate. Leaves rustle. Keep his head up. Seven seals stitch the scrotumtight sea.
harassing *acts of* *violence*	He is coming with the clouds. With a voice of whistling seawind. Gradual coming and increase. Fine china chinks. Staples and safety pins inch across tables.
limited *evacuation* *(20%)*	Pimples bubble beneath skins of paint. He's throwing up. Reports of metallic taste in mouths. From our vantage point we can see a sort of apocalypse in everyman. And climb that stairway
spectacular *show of force*	A voice behind me like a trumpet. With a cry of stormbirds. Citrus fruits burst. Radio hosts abandon posts. A kind of uniform vibration like out-of-service subways passing. Hair stands on end.

justifiable counterforce attack	Towers sway visibly in skyline. Teeth loosen. Then I turned to see the voice. Anthropomorphic clouds in southern hemisphere. Branches wave. A cemetery? Water stands take twenty-fourth street then
nuclear use threshold	changes vortex direction. Strangers attack one another. Tablecloths become reflective. Perceptible decrease in weight of objects. Amnesia and go right ahead, sir. China doorknobs shatter.
evacuation (70%)	Lobsters won't boil. Leashed dogs whine. Language fails. Hardwood floors moisten. Living will envy the metallic objects grind towards New Mexico. He's burning up. Shadow puppetry
exemplary attacks	reported on open the gateuweather radar. Ships in harbour rise, rest atop the winedark sea the sea. Systematic murder of minorities. And that was why she had to leave Skibbereen.
reciprocal reprisals	Hair falls out. Johannine. Johannine. Rape and murder according to the position he holds, the seven stars. Three-tone ringing. Towers snap. Roadkill resurrects. Everything changed. Necklaces ascend

slow-motion counter-city war	Save man's way of thinking. Weightlessness. Cover him. Feet hover and kick. Towers detach. Clarion. Clarion. Oceans stand. Universal self-mutilation. Laughter.
nuclear war is general	Eyes closed. Arms outstretched. With the moon underfoot. Curvature of earth perceptible. His jaw is clenching how I've missed you. Look there, look there. Everywhite. Everwhite. Where?

after Herman Kahn

WHERE I WAS A FLOWER OF THE MOUNTAIN
for Amy Benkard Rose

Oppenheimer turns through the wind from the driver's seat:
No you aren't you'll be fine son trust me.

Oppenheimer's tie a babbling alphabet of hieroglyphs.

Yes cicadas are perishing in their hulls of sound.
Yes field mice are dying in their narrows underground.
Yes the pink stamps of Johanna's heels which flash
Rose when she tip-toes, waving from our doorway
Are perishing, as are her borrowed hands,
Her lips, tomorrow, and the planetary tufts
Of dust that pluft up from her sandals in the sun.
Yes even the moon hangs thanatonic on its hook.

And though the night keeps going on
The insubstantial physicist, faded to a rack,
Speaks the words that would become his epitaph:

But I am yes one life is all one body only
Yes I am I am Robert yes I am yes

after Eric Ormsby

These poems appeared in ARC Magazine, The Best Canadian Poetry, Best Practices: The Scream Alumni Night Anthology, Border Crossings, Descant, Event, Guy Maddin's My Winnipeg (Coach House Books 2009), Maisonneuve, The Malahat Review, Misunderstandings Magazine, nthposition, Rhythm, and The Walrus.

Thanks are due to The Canada Council for the Arts and Le Conseil des Arts et des Lettres du Québec. And to my editor, Ken Babstock.

M. L.

MICHAEL LISTA was born in Toronto in 1983.

AUTHOR PHOTOGRAPH: JONCARLO LISTA